YOUR KNOWLEDGE HAS VALUE

- We will publish your bachelor's and master's thesis, essays and papers

- Your own eBook and book - sold worldwide in all relevant shops

- Earn money with each sale

Upload your text at www.GRIN.com and publish for free

Bibliographic information published by the German National Library:

The German National Library lists this publication in the National Bibliography; detailed bibliographic data are available on the Internet at http://dnb.dnb.de .

Imprint:

Copyright © 2016 GRIN Verlag, Open Publishing GmbH
Print and binding: Books on Demand GmbH, Norderstedt Germany
ISBN: 9783668372832

This book at GRIN:

http://www.grin.com/en/e-book/350886/proximate-and-heavy-metal-analysis-of-pumpkins-brought-to-two-different

Adekunle Jelili Olaoye

Proximate and heavy metal analysis of pumpkins brought to two different markets in Osun State Nigeria (Owode Ede and Ota-Efun Osogbo)

GRIN Publishing

PROXIMATE AND HEAVY METAL ANALYSIS OF PUMPKIN BOUGHT IN TWO DIFFERENT MARKETS IN OWODE EDE AND OTA-EFUN OSOGBO, OSUN STATE, NIGERIA.

OLAOYE, A.J

Chemistry Department, Ladoke Akintola University of Technology, Ogbomoso, Oyo State, Nigeria

Abstract

Proximate and heavy metal analysis were carried out o the leaves of pumpkin bought in two different market in Osun State Nigeria (Owode Ede and Ota-Efun Osogbo). The result of proximate analysis showed that the leaves contain; Moisture (2.88 % and 2.53%), Ash (14.45% and 13.90%), Crude Fibre (2.6 % and 2.3 %), Fat (13.20 % and 12.85%), Protein (25.66 % and 25.40 %) and Carbohydrate (41.21 % and 43.02 %) respectively for A and B. This shows that the pumpkin leaves could be used as food to supplement these nutrients. The result of heavy metal analysis using (AAS) analysis reveals that concentration of Zn (2.32 mg/g and 1.76 mg/g), Cu(1.55 mg/g and 1.05 mg/g), Fe (5.98 mg/g and 5.30 mg/g) and Mn (0.21 mg/g and 0.16 mg/g) for A and B respectively. Because of high concentration of Zn and Fe, it was concluded that pumpkin leaves could be a good source of energy and proteins for human and animals.

Key words: Proximate, Heavy metal, Pumpkin, Carbohydrate and Fibre.

Contents

1.0 INTRODUCTION

Vegetables are the edible parts of herbaceous plants that are consumed wholly or in parts, raw or cooked as part of main dish or salad; they may be aromatic, bitter or tasteless (Dhellot*et al.*, 2006). They include leaves, stems, roots, flowers, seeds, fruits, and bulbs. They are regular ingredients in the diet of the average Nigerian and provide appreciable amounts of nutritive minerals (Ajewole, 1999). Even though the bulk of their weight is water, leafy vegetables represent a veritable natural pharmacy of minerals, vitamins and phytochemicals (George, 2003).Leafy vegetable investigated in this experiment was *Telfariaoccidentalis.*

Fluted pumpkin (*Telfairiaoccidentalis*) is a creeping vegetable that spread low across the ground with lobed leaves and long twisting tendrils. It is a warm weather crop which grows well in low lands and it tolerates elevation of some few meters above the ground (Horsfall*et al*, 2005). It thrives best in soil rich in organic matters; fluted pumpkin (*TelfairiaOccidentalis*) as it is commonly known in Nigeria plays an important role in human and livestock nutrition, most especially in southern part of the country. It is source of protein, oil, fat, mineral and vitamins, which makes the leaves of the vegetable more useful in the preparation of several delicacies mostly in southern Nigeria and also other parts of the country (Aletor, *et al.,*2002).The leaves contain a high amount of vitamins A and C, antioxidants, hepatoprotective and antimicrobial properties (Oboh et al., 2006). The leaf extract is useful in the management of cholesterolemia, liver problems and impaired defense immune systems (Eseyin et al., 2005).

Fluted pumpkin thrives better in the early part of the rainy season; it also can be grown in the garden and farmed as vegetables. Fluted pumpkin has received a considerable attention in the past years; this is due to their nutritional and health protective values of the proteins from itsleaves and seeds (Achinewu,1987).Also the production of fluted pumpkin as dry season vegetable has developed from small garden to a form of commercial farming because of the production of local and exotic vegetables.(Edem et al,2009)

The root of fluted pumpkin is believed to be highly toxic to man if consumed. (Akpanunam 1984) reported on the effect of processing and temperature on its ascorbic acid and also the total carotenoids content and on its biochemical composition, (Achinewu 1987), also report on the thiamine riboflavin and niacin content of the fermented fluted pumpkin while (Akpanunam,1998) evaluated it's mineral and toxicant content, (Ajayi et al 1980), and (Apkanunam,1984) also reported on the loss of vitamin-c content in Cooked *telfairiaoccidentalis.*

1.1 USES OF TELFAIRAOCCIDENTALIS

The fruits of the plant are large and inedible buthe seeds contain up to 30% protein and can be boiledand eaten, or ground into powder for soup. The seeds ofthe plant can also be fermented for several days andeaten as a slurry(Badifu and Ogunsina,1991). The Leaves arealso rich in essential and non-essential amino acids, vitamins and minerals (Fasuyi, 2006).Theherbal preparation of the plant has been employedin the treatment of suddenattack of convulsion, malariaand anaemia (Gbile, 1986). Based on its use as ahaematinic, its effects on haematological indices hadbeen scientifically investigated and reported. The dietpreparation of the air-dried leaves of the plant significantlyincreased red blood cell count, white blood cellcount, packed cell volume and haemoglobin concentrationin rats while the dietary preparationmade with the sun-dried leaves had no significant effecton haematological parameters in birds (Alada, 2000).

1.2 BLOOD PRESSURE REDUCTION

Blood pressure control is important for the prevention of heart diseases and stroke which can be influence by numerous factors, hypertension can be caused by atherosclerosis, in balance in the rennin-angiotention system and hyperinsulinemia which increase sodium retention in the body. Consequently, a general nutritional plan to minimize hypertension risk include attaining and maintaining healthy body weight, consuming a diet rich in calcium, phosphorus and magnesium; Also animal products with the vegetable diet has been found to reduces blood pressure in normotensive and hypertensive Individuals. Lower intake of fat and higher intake of dietary fiber, minerals, and vegetable diet is believed to reduce blood pressure. (Aletor *,et al.,2002*).

2.0 MATERIALS AND METHODOLOGY

2.1 EQUIPMENT AND APPARATUS

- Fume cupboard

- Hot plate

- Micro digestion flask

- Volumetric flask

- Glass funnel

2.2 REAGENT S

10-20ml sample

40% of 10ml -20ml sodium hydroxide solution

2% of 5ml Boric Acid

0.1M HCl or H_2SO_4(For titration)

Indicator(methyl Orange)

Perchloric acid

Nitric acid

Digestion tablet

Distilled water

2.3 SAMPLING

The species of fluted pumpkin were randomly obtained from Owode market, Edeand Ota-Efun market, Osogbo both in Osun State. The samples were air dried for some weeks and after it has been dried, they were pounded using mortar and pestle, then sieve and stored in an air-proof polyethylene bags in the laboratory until when needed for analysis.

2.4 EXPERIMENT

Moisture content determination

This was done according to AOAC (1990), where 2g of the fresh leaves of pumpkin were weighed (W_1) into pre-weighed crucible (W_0) and placed into a hot drying oven at $105^{o}C$ for 24 hours. The crucible was removed, cooled in a desiccators and re-weighed (W_2). The processes of drying, cooling and weighing were repeated until a constant weight (W_2) was obtained. The weight loss due to moisture was obtained by the equation.

% Moisture $= \dfrac{w1-w2}{2g}$ x 100 %

Crude Protein Content Determination

Crude protein was determined using the micro-Kjeldhal method as described by AOAC (1990), whereby 2g of sample was weighed along with 20cm3 of distilled water into a micro Kjeldahl digestion flask. It was shaken and allowed to stand for some time. One tablet of selenium catalyst was added followed by addition of $20cm^3$ conc. H_2SO_4. The flask was heated on the digestion block at 1000C for 4 hours until the digest became clear. The flask was removed from the block and allowed to cool. The content was transferred into a $50cm^3$ volumetric flask and diluted to the mark with water. An aliquot of the digest ($10cm^3$) was transferred into another Micro-Kjeldhal flask along with $20cm^3$ of distilled water and placed in the distilling outlet of the Micro-Kjeldhal distillation unit. A conical flask containing $20cm^3$ of boric acid indicator (2% boric acid containing bromo cresol green-methyl red indicator) was placed under the condenser outlet. Sodium hydroxide solution ($20cm^3$, 40%) was added to the content in the Kjeldhal flask by opening the funnel stop cock. The distillation started and the heat supplied was regulated to avoid sucking back. When all the available distillate was collected in $20cm^3$ of boric acid, the distillation was stopped. The nitrogen in the distillate was determined by titrating with 0.01M of H_2SO_4. . The crude protein was calculated using the equation; % crude protein = % N x 6.25. The nitrogen content of the sample is given by the formula;

$$\% \text{ Protein} = \frac{TV \times Na \times 0.014 \times V_1}{G \times V_2} \times 100 \%$$

TV = Titre value of acid (cm^3)

Na = Concentration or normality of acid

V_1 = Volume of distilled water used for distilling the digest ($50 \ cm^3$)

V_2 = Volume of aliquot used for distilling the digest ($10 \ cm^3$)

Ash Content Determination

This was done according to AOAC (1990), where 2g of the powdered leaves sample was weighed (W_1) into pre-weighed empty crucibles (W_0) and place into a muffled furnace at 600^oC for 3 hours. The ash was cooled in a desiccator and weighed (W_2). The weight of the ash was determined by the difference between the powdered leaves sample, pre-weighed crucible and the ash in the crucible.

Percentage ash was calculated as:

$$\% \text{ Ash} = \frac{W_2 - W_1}{2g} \times 100\,\%$$

Fatty Material Content Determination

Two (2) grams of the pumpkin leaves sample was placed on a filter paper and put into the soxhlet extractor and extracted into a pre-weighed round bottom flask with low boiling petroleum ether (40 – 600C) using a soxhlet extractor for 8 hours. The solvent was recovered by rotary evaporation, and drying was completed in a freeze dryer. Finally the flask and its content were heated at 900c in an oven for 2 hours, and cooled in a dissector and weighed. The process of heating and cooling was repeated until a constant weight was obtained. AOAC (1990)

$$\% \text{ Fatty materials} = \frac{\text{Weight of fatty material} \times 100\,\%}{2\,g}$$

Crude Fibre Content Determination

Percentage crude fibre was determined by the method described in AOAC (1990) in which 2g of ground sample was weighed (W_0) into a $1dm^3$ conical flask. Water ($100cm^3$) and $20cm^3$ of 20% H_2SO_4 were added and boiled gently for 30 minutes. The content was filtered through Whatman No.1 filter paper. The residue was scrapped back into the flask with a spatula and $100cm^3$ of water and $20cm^3$ of 10% NaOH were added and allowed to boil gently for 30 minutes. The content was filtered and residue was washed thoroughly with hot distilled water, rinsed once with 10% HCl and twice with ethanol and finally 3 times with petroleum ether. It was allowed to dry and scrapped into the crucible and dried over night at 1050c in an air oven. It was then removed and cooled in a desiccator. The sample was weighed (W_1) and ashed at $600^{\circ}C$ for 90 minutes in a muffled furnace. It was finally cooled in a dessicator and weighed again (W_2). The percentage crude fibre was calculated using equation.

$$\% \text{ Crude fibre} = \frac{W_1 - W_2}{W_0} \times 100\,\%$$

Carbohydrate Content Determination

The method of AOAC (1990) was adopted where the total proportion of carbohydrate in the leaves sample was calculated by subtracting the % sum of food nutrients: % protein, % crude fibre and % ash from 100%, as

% CHO = 100% - (% crude protein + % crude fibre + % ash + crude lipid).

3.0 RESULT AND DISCUSSION OF RESULT

TABLE 1; CONCENTRATION OF SOME HEAVY METALS IN FLUTED PUMPKIN

Samples (mg/g)	Cu	Fe	Mn
A	1.55 ±0.005	5.98 ±0.007	0.21 ± 0.0006
B	1.05± 0.005	5.30 ±0.007	0.16 ±0.0006

TABLE 2 PROXIMATE COMPOSITION OF PUMPKIN

Sample	Moisture	Protein	Ash	Fat	Carbohydrate	Fibre
A	2.88±0.0002	25.66±0.006	14.45±0.003	13.20±0.0007	41.21 ± 0.005	2.60±0.006
B	2.53±0.0002	25.40±0.006	13.90±0.003	12.85±0.0007	43.02 ± 0.005	2.30 ± 0.006

3.1 RESULT

The result shows that the concentration in (*Telfairiaoccidentalis*)leaves of Zinc (Zn) ranged from 2.32 mg/g for A to 1.76 mg/g for B, the concentration of copper (Cu) ranges from 1.55 mg/g for Ato 1.05 mg/g for B, the concentration of Iron (Fe) ranged from 5.98 mg/g for A to 5.30mg/g for B while concentration of manganese (Mn) ranged from 0.21mg/g for A to 0.16 mg/g for B respectively.

The result of proximate composition of pumpkin (*Telfairiaoccidentalis*) ranges from moisture 2.88 for A to 2.53 for B, the protein content range from 25.66 for A to 25.40 for B, the ash content varies from 14.45 for A to 13.90 for B, the fat content ranges from 13.20 to 12.85 for A and B the fibre content varies from 2.6 to 2.3 for A and B respectively while the carbohydrate content varies from 41.21 for A to 43.02 for B respectively.

3.2 DISCUSSION

The result of the experiment revealed that the concentration of heavy metals in each of the sample varied from one market to the other. It was showed that concentration of market A is differ from that of market B because of environmental factors of Owode Ede which differ from that of Osogbo.

The moisture content of both sample were range from (2.88 of A to 2.33 B). This varies significantly with those obtained by (Ogunlade *et al.*, 2011, Amoo and Agunbiade ,2009 and Akoja and Amoo2011) respectively for Guinea peanut, *Pterygotamacrocarpa, Hexalobuscrispiflorus*and*Clitandratolana*. However, the result of moisture of both sample were very close to what obtained by (Ogbe and John 2012), for Moringaoleifera leaves but extremely lower than the what obtained by (Blessing *et al.*,2011). Low amount of moisture in crops makes them not to vulnerable to microbial attack, hence, spoilage (Desai and Salunkhe, 1991).The moisture content is low compare to shoot of BorassusAethiopumart, xanthosemsagihifolumGnetumbuchhoysianium and Adansoniadigitatarespectevily as reported by (Gafar *et al.*, 2009).The low moisture content of pumpkin found in this work is an indication that is to a large extent protected from enzymatic decomposition (Fai *et al.*, 2013).

The difference in % moisture is as a result of different in environmental condition.The ash content of the leaf indicates that the leaves contain high level of ash compared to most of the leafy vegetable. It is lower than Balsam and higher than Indegoferaastragalina and sweet potato leaves Ladan et al., 1996, and pumpkin reported by (Fai *et al.*, 2013).The high level of ash indicates that pumpkin are rich in mineral element. The fat content of this study is higher compared to spinarch leave and Amarantushybridus leave and (Nwaogu *et al.*, 2000) and water spinarch reported by (Hassan and Umar 2004). It was also higher than the result obtained by *(Fai et al.*, 2013) onpumpkin. Fats are the principal source of energy therefore; pumpkin can be used as source of energy.

 The protein content of obtained in this studied is higher compared to what obtained by previous workers (Nwaogu *et al.*, 2000) on water spinach, (Fai *et al.*, 2013) on pumpkin and almost the same as what obtained by (Hassan and Umar 2004) on sweet potato leaves. Pumpkin is a source of protein compared to other leafy vegetables (Salahu, 2005).

The fibre content of pumpkin in this studied is almost the same as what obtained by (Fai et al., 2013) on pumpkin, *(Gafar et al.*, 2009) on Indigoferaastragalina but lower compared to what obtained by *(Gafar et al.*, 2009) on sweet potatoes leaves. But the value is within the reported literature which ranges from 0.70 – 12.00% for most leafy vegetables (Hassan and Umar 2004). Since the leaves contain crude fibre, it could help to reduce the level of cholesterol in the body and also aid ingestion. It could also help to reduce the level of glucose in the body (Fai *et al.*, 2013).

The carbohydrate content of the leaves in this studied is considerably low compared to previous worker (Fai *et al.*, 2013) on pumpkin, (Hassan *et al.*, 2006) on water spinach and Tribalusterrestris.Carbohydrate is a principal source of energy. Therefore the pumpkin can be served as an alternative source of carbohydrate for the consumers (Fai *et al.*, 2013) on pumpkin.

CONCLUSION

The result of the study shows that the heavy metal concentration in the studied fluted pumpkin *(T.Occidentalis)*was within the acceptable limit for human consumption. The high concentration of zinc and Iron is an indication of the nutritional and health influence of fluted pumpkin to man and society in general. Likewise, the proximate analysis of fluted pumpkin shows that it is a source of both protein and carbohydrate, therefore it is a good food supplement for human consumption.

References

Dhellot, J.R., Matouba, E., Maloumbi, M.G., Nzikou, J.M., Safou-Ngoma, D.G., Linder, M., Desobry, S., Parmentier, M. (2006). Extraction, chemical composition and Nutritional characterization of vegetable oils: Case of Amaranthushybridus (Var 1 and 2) of Congo Brazzaville. *Africa. Journal of Biotechnology.* 5 (11):1095-1101.

Ajewole, K. (1999). Analysis of the nutritive elements in some Nigerian leafy vegetables.Proceedings of 23rd Annual Nigerian Institute of Food Science and Technology (NIFST) Conference 25-27 October, Abuja, Nigeria.6-8

George, P.M. (2003). Compositional Analysis Method in Manuals of Food Quality Control. Encyclopedia of Foods 7(1):203-232

Horsfall, M. and Spiff I.A. (2005). Equilibrium sorption study of Al, CO and Ag in aqueous solution by fluted pumpkin (*telfairia occidentalis*). Waste biomas. Acta chin slov,52;174-181

Aletor .O., Oshodi, A.A. and Ipinmoroti, K. (2002); Chemical composition of common leafy vegetables and functional properties of their leafy protein concentrates .*Food chemistry* 78;63-68

Association of Official Analytical Chemists, (1990). Official methods of analysis, 16th ed., Washington DC, U.S.A

Oboh, G., Nwanna, E.E.and Elusiyan, C.A. (2006). Antioxidant and antimicrobial properties of *Telfairia occidentalis* (Fluted pumpkin) leaf extracts. Journal of Pharmacology and Toxicology 1:167–175.

Eseyin, O.A., Igboasoiyi, A.C., Oforah, E., Ching, P.and Okoli. B. C. (2005). Effects of leaf extract of *Telfairiaoccidentalis* on some biochemical parameters in rats. *Global Journalof Pure and Applied Science*11:17-19.

Achinewu, S.C. (1987); Thiamin riboflavin and niacin content of fermented fluted pumpkin (*T.Occidentalis*) castos seeds (*ricinus communis*) Nig*eria Nutritional science, 8; 51-61*

Edem, C.,Dosumu,A., Miranda, I. and Bassey, F. I.(2009). Distribution of heavy metals in leaves, stem and Roots of Fluted pumpkin (*telfairia occidentalis*). *Pakistan Journalof Nutrition* 8 (3): 222-224,

Akpanunam, M.A. (1984). Effects of wilting, blancing and storage temperature on ascorbic acid and total carotenoids content of some Nigeria fresh vegetables plant food.*Human*

Nutriton,34;177-180

Ajayi S.O, Oderinde S.F and Osibanjo (1980). Vitamin-C losses in cooked fresh leafy vegetables.*Food Chemistry* 5;243-247

Badifu, G.I.O and Ogunsua, A.O. (1991).Chemical composition of kernels from some species of Cucurbitacea grown in Nigeria. *Plant food human nutrition,*41;35-44

Fasuyi, A.O. (2006). Nutritional potentials of some tropical vegetable leaf meals; chemical characterization and functional properties. *African journal. Biotechnology* 5; 49-53

Gbile, Z.O. (1986). Ethnobotany,taxonomy,and conservation of medial plants in the state of medicinal plant research in Nigeria. Edited by sofowora A.O p.19

Alada, A. (2000). The heamatological effects of *Telfairia occidentalis* diet preparation: *African Journal Burned Research*: 3(3);185-186

Ogunlade, I., Ilugbiyin, A. and Osasona, I.A., (2011). A Comparative Study of Proximate Composition, Anti-Nutrient Composition and Functional Properties of *Pachiraglabra*and *Afzelia Africana seed flours. African Journal Food Science.* 5(1)32-35.

Amoo, I.A. and Agunbiade, F.O., (2009). Some Nutrient and Anti-Nutrient Components of *Pterygotamacrocarpa.The Pacific Journal of Science and Technology.* 10(2): 949-955.

Akoja, S.S. and Amoo I.A., (2011). Proximate Composition of Some Under-Exploited Leguminous Crop Seeds. *Pakistan Journal of Nutrition.* 10 (2): 143-146.

Ogbe, A.O. and John, P.A., (2012). Proximate study, Mineral and Anti-nutrient composition of *Moringaoleifera*leaves harvested from Lafia, Nigeria: Potential benefits in poultry nutrition and health. *Journal of Microbiology, Biotechnology and Food Sciences.* 1 (3): 296-308.

Blessing, A.C., Ifeanyi, U.M. and Chijioke, O.B., (2011).Nutritional Evaluation of Some Nigerian Pumpkins (*Cucurbita spp.*).Fruit, Vegetable and Cereal. *Science and Biotechnology.Global Science Books. 5 (2): 64-71.*

Desai, B.B. and Salunkhe, D.K., (1991). Fruits and Vegetables. In Foods of Plant Origin. Production, Technology and Human Nutrition, AVI, *New York,*301-355.

Gafar R. A., Abdulrahman, A. R.Mahmod, N.Z and Vasudevan R.(2009). Proximate Analysis ofDragon Fruit (Hylecteruspolyhizus). *American Journal ofApplied Sciences* 6(7):1341.

Fai, F.Y., Danbature, W.L., Auwal,Y and Usman, Y.M. (2013). Proximate and some mineral analysis of Pumpkin leaf. *Journal of physical Science and Environmental safety*, (3) 1-8

Ladan, M. J. Bilbis, L. S. and Lawan, M.(1996). Nutrient Composition ofSome Green Leafy VegetablesConsumed in Sokoto, Nigeria.*Journal of Basic and AppliedSciences*, 5(1 and 2):39-44.

Nwaogu, L. A., Ujowundu C. O., and,Mgbemena, A. I. (2000). Studieson the, Nutritional, and PhytochemicalComposition ofAmarantus hybridus Leaves.*Bio-Research* 4(1):28-31.

Hassan, L. G. and Umar, K. J. (2004).Anti-nutritional Factors inTribulus terrestris (Linnaeus) Leaves and Predicted Calciumand Zinc Bioavailability.*Nigerian Journal of AppliedSciences,* 5(2 and 3):57-61.

Salahu S.O. (2005). Phyto-chemicalScreening, Nutrient and, Antinutrient Composition of Selected Tropical Green Leafy Vegetables. *African Journal ofBiotechnology,* 4:497-501.

Hassan, L.G., Umar K. J., Daigogo, S. M.and Ladan, M.J. (2006). Protein and, AminoAcids Compositionof African Locust Bean (Parkiabiglobosa L). *Journal of Tropicaland Subtropical Agro system 11- 32*

YOUR KNOWLEDGE HAS VALUE

- We will publish your bachelor's and
  master's thesis, essays and papers

- Your own eBook and book -
  sold worldwide in all relevant shops

- Earn money with each sale

Upload your text at www.GRIN.com
and publish for free